YOUR KNOWLEDGE HAS VALUE

- We will publish your bachelor's and
  master's thesis, essays and papers

- Your own eBook and book -
  sold worldwide in all relevant shops

- Earn money with each sale

Upload your text at www.GRIN.com
and publish for free

Chayan Mehta, Arpita Singh, Rahul Lohiya, Surabhi Bhattacharya

Analysis of Harmonics Injected by Single Phase Inverter

GRIN Verlag

Imprint:

Copyright © 2013 GRIN Verlag GmbH
Druck und Bindung: Books on Demand GmbH, Norderstedt Germany
ISBN: 978-3-656-54857-7

This book at GRIN:

http://www.grin.com/en/e-book/265312/analysis-of-harmonics-injected-by-single-
phase-inverter

A

Minor Project Report

On

Analysis of Harmonics Injected by Single Phase Inverter

Submitted in partial fulfillment of the requirements of the degree of

Bachelor of Technology

In

Electronics & Communication Engineering

Arpita Singh

Chayan Mehta

Rahul Lohiya

Surabhi Bhattacharya

SIR PADAMPAT SINGHANIA UNIVERSITY

UDAIPUR

CERTIFICATE

This is to certify that the Minor/Major Project entitled *'Analysis of Harmonics Injected by Single Phase Inverter'* being submitted by Arpita Singh, Chayan Mehta Rahul Lohiya ,Surabhi Bhattacharya in fulfillment of the requirement for the award of degree of Bachelor of Technology in Discipline of Electronics & Communication Engineering, has been carried out under my supervision and guidance. The matter embodied in this thesis has not been submitted, in part or in full, to any other university or institute for the award of any degree, diploma or certificate.

Prof. Ritesh Tirole
Assistant Professor

Prof. Udayprakash Raghunath Singh
Head of Department
Electronics and Communication Engineering

Dr. Achintya Choudhury
Dean, School of Engineering
Sir Padampat Singhania University
Udaipur- 313601 Rajasthan India

Acknowledgement

We would like to express the deepest appreciation to our Mentor Mr. Ritesh Tirole, who has the attitude and the substance of a genius: he continually and convincingly conveyed a spirit of adventure in regard to research and scholarship, and an excitement in regard to teaching. Without his guidance and persistent help this dissertation would not have been possible.

We would also like to thank all my instructors and teachers, who throughout my educational career have supported and encouraged me to believe in my abilities. They have directed me through various situations, allowing me to reach this accomplishment.

We deeply express our sincere thanks to our Head of Department Mr. Udayprakash Raghunath Singh for encouraging and allowing us to do the project on the topic "Analysis of Harmonics Injected by Single Phase Inverter".

We take this opportunity to thank all our lecturers who have directly or indirectly helped our project. We pay our respects and love to our parents and all other family members and friends for their love and encouragement throughout our career. Last but not the least we express our thanks to our friends for their cooperation and support.

Team Members

Arpita Singh

Chayan Mehta

Rahul Lohiya

Surabhi Bhattacharya

Abstract

The power electronics device which converts DC power to AC power at required output voltage and frequency level is known as inverter. As we have found that different inverters are used for different equipment's so in our project titled
" Analysis of Harmonics injected by Single phase Inverter" , we are analyzing the harmonics present in single phase voltage source inverter using different loads (R,RL and RLC).

We are analyzing Harmonics using MATLAB tools like Scope for harmonics and Simulink powergui for analysis of FFT of different Signals.

Table of Contents

Chapter 1 Introduction..**6**

   Software Used...7

      Single Phase Voltage Inverter 8

      Single Phase Half Bridge Inverter..8

      Single Phase Full Bridge Inverter...9

Chapter 2 Analysis of Single Phase Inverter.........................**10**

   1. Analysis of Single Phase Half Bridge Inverter with R load............................11

      Single Phase Inverter Half Bridge with RL load 12

      Single Phase Inverter Half Bridge with RLC Load.......................14

   2. Analysis of Single Phase Full- Wave Inverter with R load 16

      Single Phase Inverter Half Wave with RL load 18

      Single Phase Inverter Half Wave with RLC load.......................20

Chapter 3 Results..**22**

   1. FFT Analysis of Single Phase Half Bridge Inverter with R load 23

      Single Phase Inverter Half Bridge with RL load 24

      Single Phase Inverter Half Bridge with RLC Load.......................25

   2. FFT Analysis of Single Phase Full- Wave Inverter with R load 26

      Single Phase Inverter Half Wave with RL load 27

      Single Phase Inverter Half Wave with RLC load.......................28

Chapter 4 Conclusion **29**

Chapter 5 Further Scope **30**

References

Chapter 1

Introduction

A Circuit that converts dc power into ac power at desired output voltage and frequency is called an inverter. Some industrial applications of inverters are for adjustable-speed ac drives, induction heating, stand by air-craft power supplies, UPS (uninterruptible power supplies).

The dc power input to the inverter is obtained from an existing power supply network or from a rotating alternator through a rectifier or a battery, fuel cell, photovoltaic array or magnetic hydrodynamic generator. The rectification is carried out by standard diodes or thyristor converter circuits.

Inverters can be broadly classified into two types:

> - Voltage Source Inverters →dc source has small or negligible internal impedance.
> - Current Source Inverters→adjustable current from a dc source of high internal impedance.

From the viewpoint of connection of semiconductor devices, inverters are classified as under:

> - Bridge inverters.
> - Series inverters.
> - Parallel inverters.

Harmonics: A harmonic of a wave is a component frequency of the signal that is an integer multiple of the fundamental frequency, i.e. if the fundamental frequency is f, the harmonics have frequencies 2f, 3f, 4f, . . . etc. The harmonics have the property that they are all periodic at the fundamental frequency, therefore the sum of harmonics is also periodic at that frequency.

Software Used: MATLAB R2010a

Tools Used: MATLAB Simulink

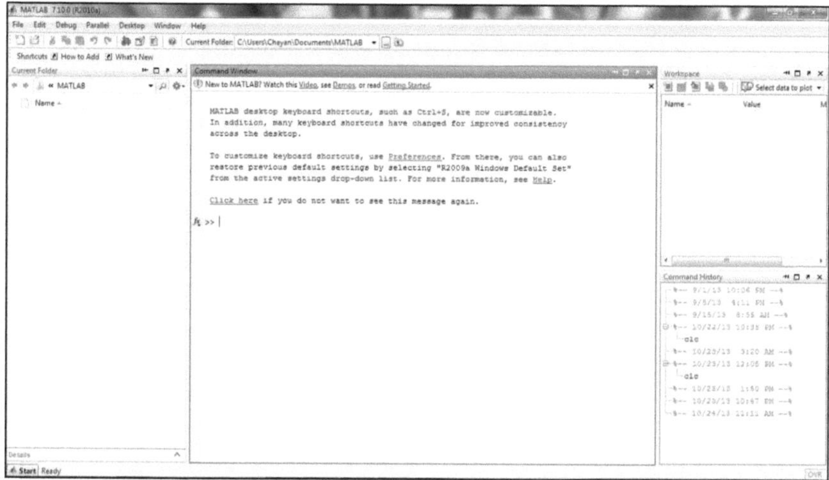

<u>Figure 1.1</u>

MATLAB Simulink:

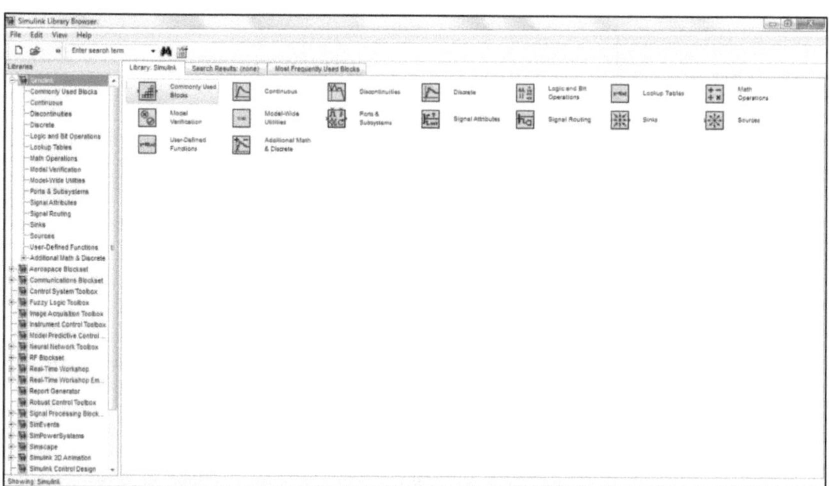

<u>Figure 1.2</u>

Single-Phase Voltage Source Inverters → Single-phase bridge inverters are of two types, namely

> Single-phase half-bridge inverters.
> Single-phase full-bridge inverters.

Single phase half-bridge inverter:

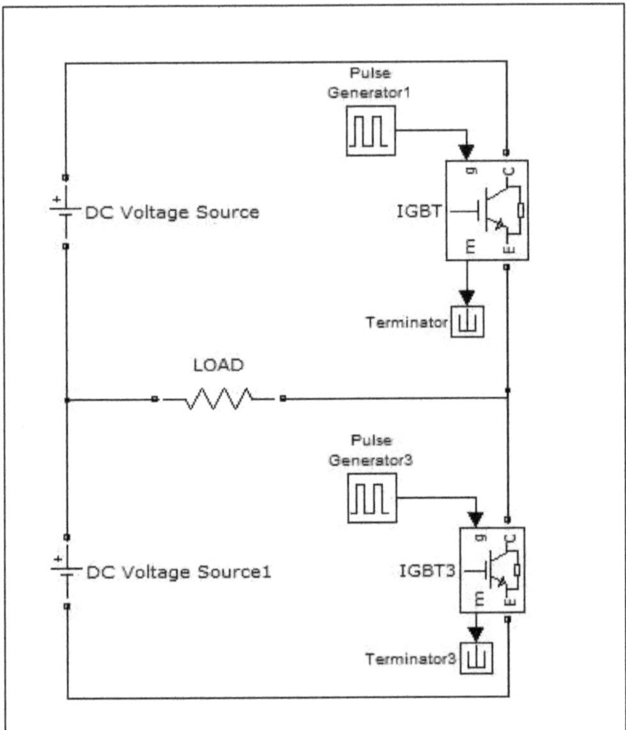

Figure 1.3 Single phase half-bridge inverter

> The circuit consist of two IGBTs, 3-wire dc supply and Pulse generators.
> The gating signal for both IGBTs are provided by pulse generators.
> Each provides opposite polarity of Vs/2 across the load.

Single phase full -bridge inverters:

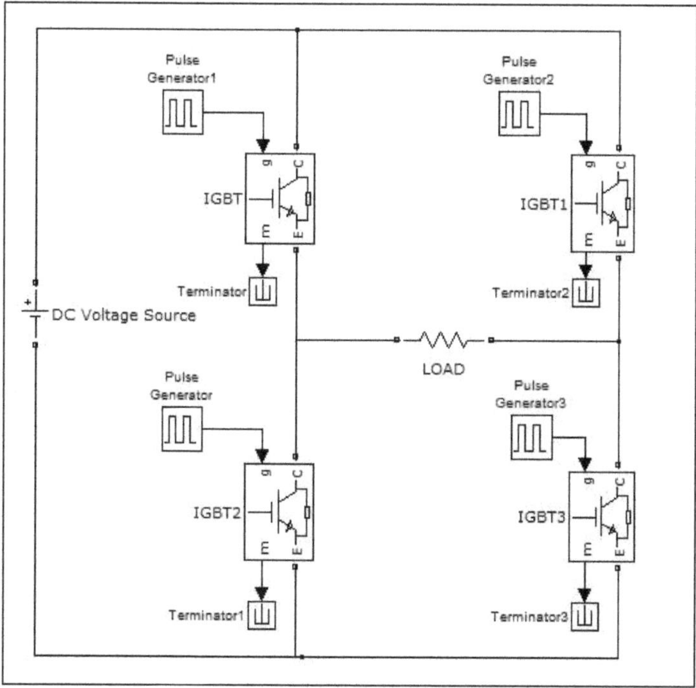

<u>Figure 1.3 Single phase full bridge inverters</u>

- ➢ The circuit consist of four IGBTs, dc voltage supply. Pulse generators and load.
- ➢ IGBT-IGBT$_3$ and IGBT$_1$-IGBT$_2$ switched on and off alternately
- ➢ The gating signal for IGBT's are provided by pulse generators.
- ➢ Each pair provide opposite polarity of V_s across the load
- ➢ The main drawback of half bridge inverter is that it requires 3-wire dc supply. This difficulty is removed by the use of full bridge inverters.

Analysis of Single-phase Bridge Inverters

The output voltage obtained from inverters is not a sine wave. It consists of fundamental component plus certain harmonics. The harmonics in the output voltage waveform of inverters leads to poor load performance and reduced system efficiency. Lower the harmonic content in the output waveform, better is the quality of the inverter.

1. Single-phase half bridge inverter

> Single-phase half bridge inverter with R load

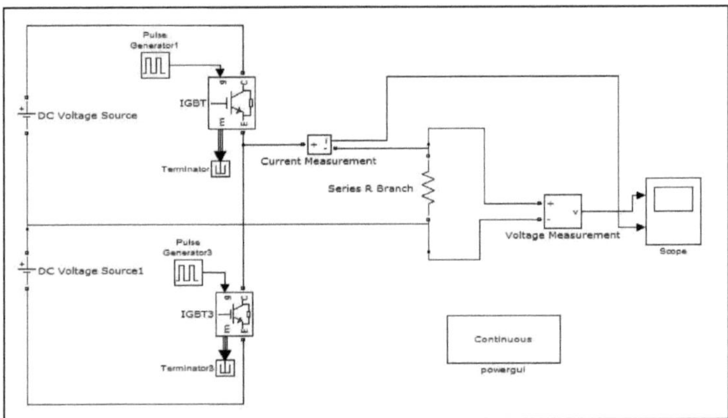

Figure 2.1 Single-phase half bridge inverter with R load

For above figure (1.3) Pulse Generator1 has following values:

Period (sec) =0.02(20ms).

Pulse width (% of Period) =50 and

Phase Delay (sec) =0. Phase Delay for Pulse Generator is 0.01sec

- In this circuit ,3-wire dc supply is used as input voltage supply.IGBTs are responsible for switching activity and the gating pulse from the pulse generators is responsible for the firing of IGBTs.

- Output voltage and Current is being measured at load using voltage and current measuring devices respectively.

- The Output Voltage and Current Waveform of Single phase half-bridge inverter with R Load are as follows:

The Load current waveform is identical with load voltage waveform →

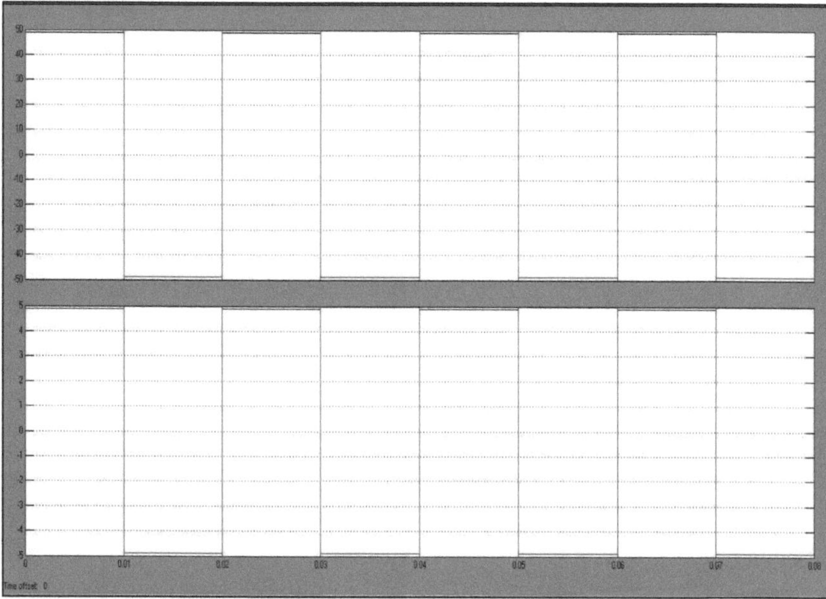

Figure 2.2 *load voltage waveform of* Single-phase half bridge inverter with R load

For Figure 2.2,

X-axis defines Time Period and Y-axis defines Voltage Value and Current Value respectively.

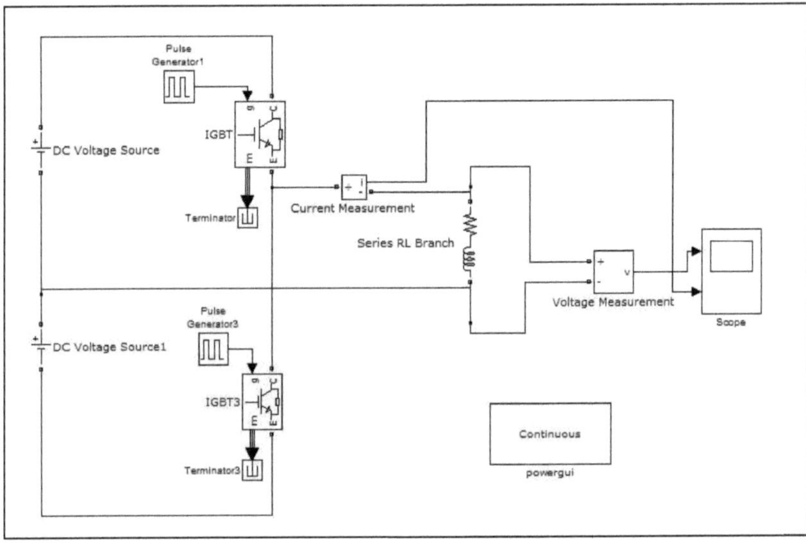

Figure 2.3 Single phase half-bridge inverter with RL Load

For above figure (2.5) Pulse Generator1 has following values:

Period (sec) =0.02(20ms).

Pulse width (% of Period) =50 and

Phase Delay (sec) =0. Phase Delay for Pulse Generator is 0.01sec

- In this circuit ,3-wire dc supply is used as input voltage supply.IGBTs are responsible for switching activity and the gating pulse from the pulse generators is responsible for the firing of IGBTs.
- Output voltage and Current is being measured at load using voltage and current measuring devices respectively.

- The Output Voltage and Current Waveform of Single phase half-bridge inverter with RLC Load are as follows:

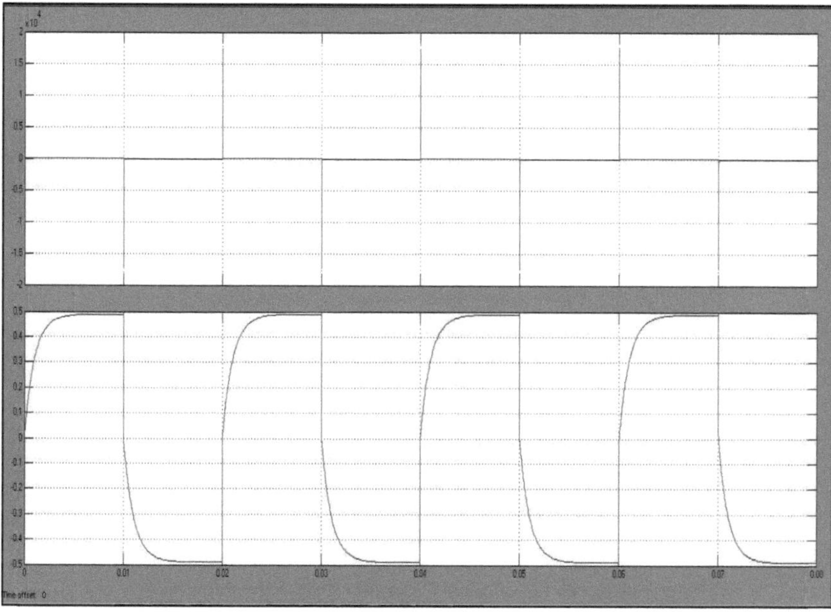

Figure 2.4

- For Figure 2.4,

X-axis defines Time Period and Y-axis defines Voltage Value and Current Value respectively.

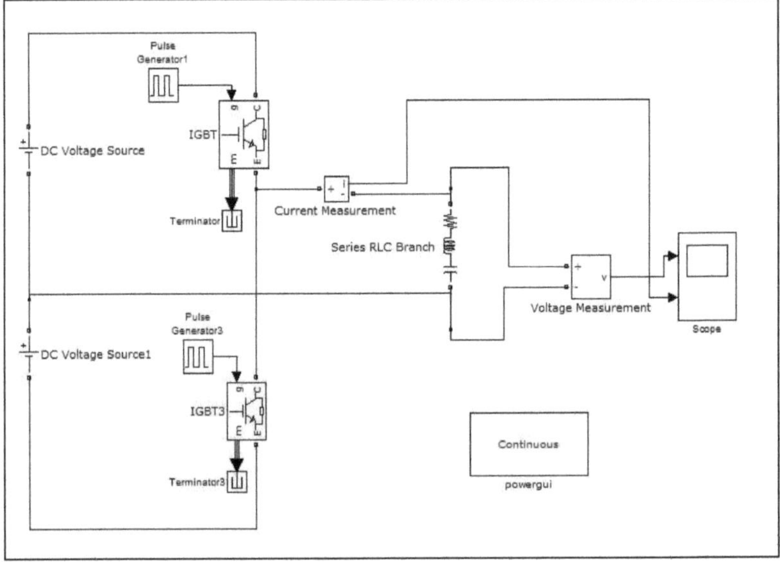

Figure 2. 5 *Single phase half-bridge inverter with RLC Load*

For above figure (2.5) Pulse Generator1 has following values:

Period (sec) =0.02(20ms).

Pulse width (% of Period) =50 and

Phase Delay (sec) =0. Phase Delay for Pulse Generator is 0.01sec

- In this circuit ,3-wire dc supply is used as input voltage supply.IGBTs are responsible for switching activity and the gating pulse from the pulse generators is responsible for the firing of IGBTs.
- Output voltage and Current is being measured at load using voltage and current measuring devices respectively.

- The Output Voltage and Current Waveform of Single phase half-bridge inverter with RLC Load are as follows:

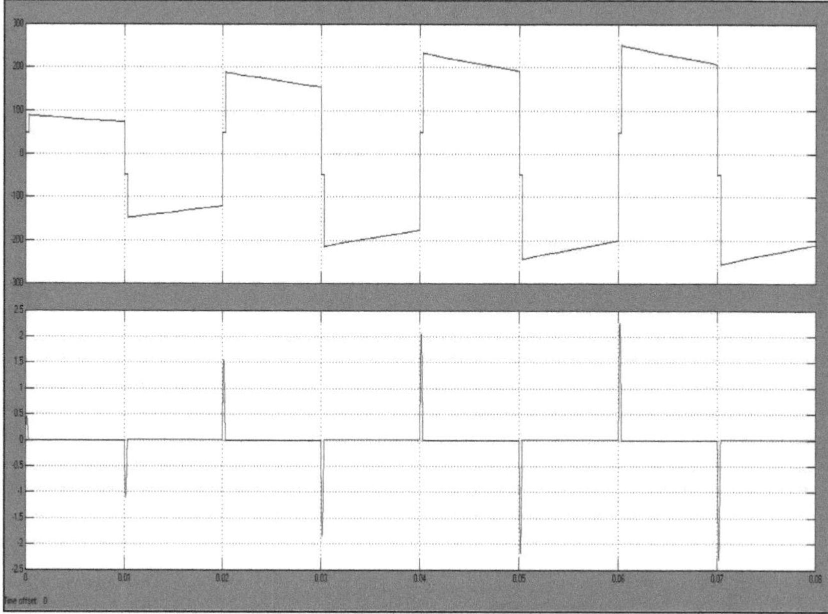

Figure 2.6

- For Figure 2.6,

X-axis defines Time Period and Y-axis defines Voltage Value and Current Value respectively.

2. Single-Phase Full bridge Inverter

> Single-Phase Full bridge Inverter with R load

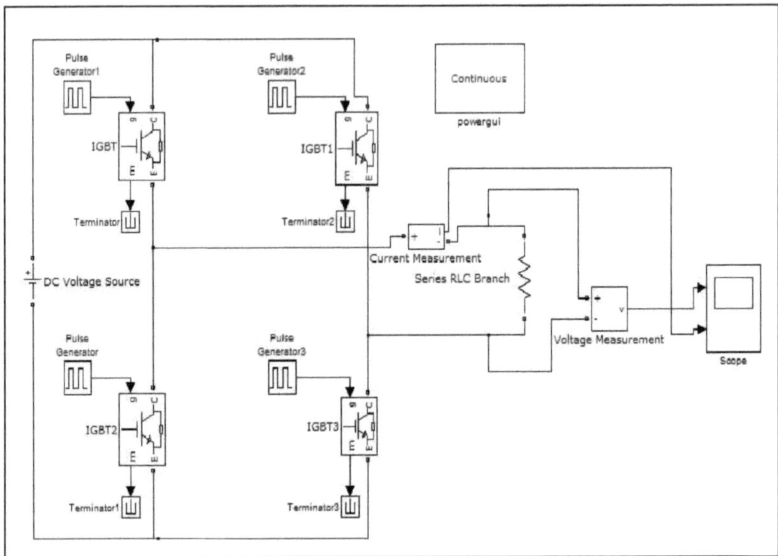

Figure 2.7 Single-Phase Full bridge Inverter with R load

For above figure (2.7) Pulse Generator1 & Pulse Generator3 has following values:

Period (sec) =0.02(20ms).

Pulse width (% of Period) =50 and

Phase Delay (sec) = 0.

Pulse Generator and Pulse Generator2 has Phase delay of 0.01sec

- In this circuit ,dc voltage source is used as input supply.IGBTs are responsible for switching activity and the gating pulse from the pulse generators is responsible for the firing of IGBTs.

- Output voltage and Current is being measured at load using voltage and current measuring devices respectively.

- The Output Voltage and Current Waveform of Single phase full-bridge inverter with R Load are as follows:

Figure 2.8

- For Figure 2.8,

X-axis defines Time Period and Y-axis defines Voltage Value and Current Value respectively.

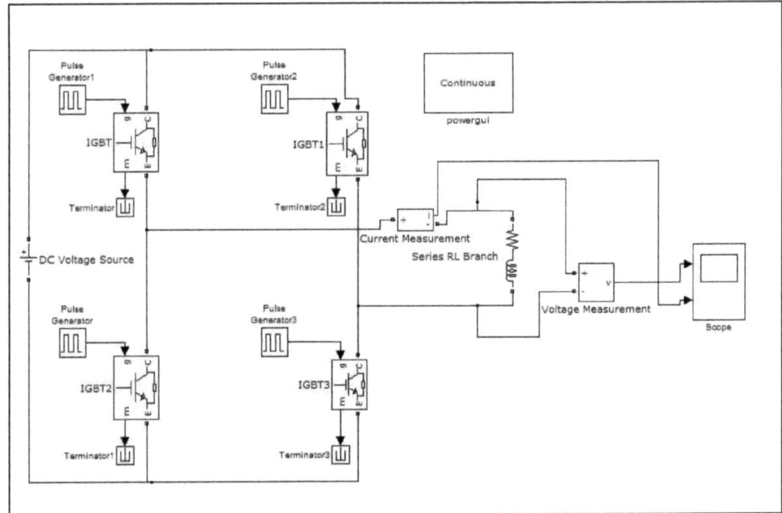

<u>Figure 2.9 Single phase full-bridge inverter with RL Load</u>

For above figure (2.9) Pulse Generator1 & Pulse Generator3 has following values:

Period (sec) =0.02(20ms).

Pulse width (% of Period) =50 and

Phase Delay (sec) = 0.

Pulse Generator and Pulse Generator2 has Phase delay of 0.01sec

- In this circuit ,dc voltage source is used as input supply.IGBTs are responsible for switching activity and the gating pulse from the pulse generators is responsible for the firing of IGBTs.
- Output voltage and Current is being measured at load using voltage and current measuring devices respectively.

- The Output Voltage and Current Waveform of Single phase full-bridge inverter with RL Load are as follows:

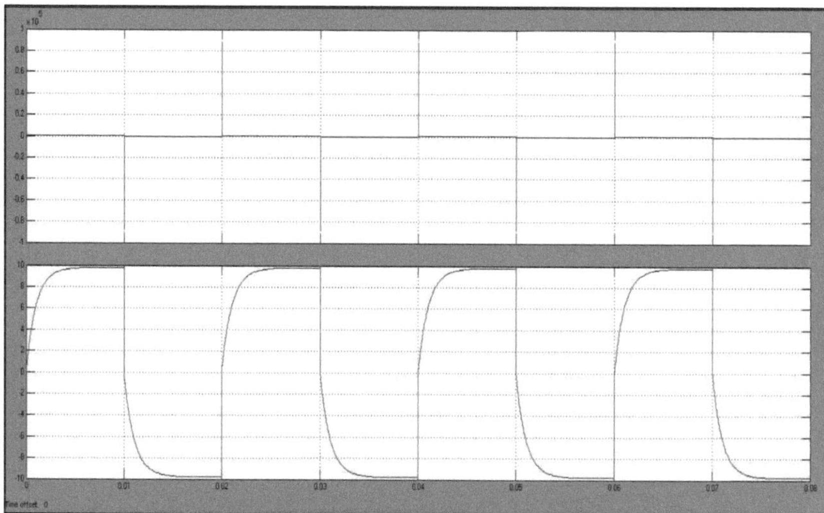

Figure 2.10

- For Figure 2.10,

X-axis defines Time Period and Y-axis defines Voltage Value and Current Value respectively

➢ Single phase full-bridge inverter with RLC Load

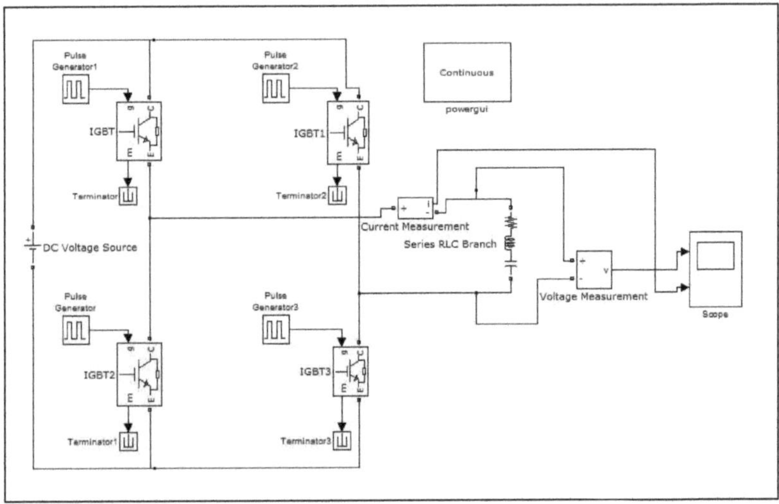

Figure 2.4 Single phase full-bridge inverter with RLC Load

For above figure (2.9) Pulse Generator1 & Pulse Generator3 has following values:

Period (sec) =0.02(20ms).

Pulse width (% of Period) =50 and

Phase Delay (sec) = 0.

Pulse Generator and Pulse Generator2 has Phase delay of 0.01sec

- In this circuit ,dc voltage source is used as input supply.IGBTs are responsible for switching activity and the gating pulse from the pulse generators is responsible for the firing of IGBTs.
- Output voltage and Current is being measured at load using voltage and current measuring devices respectively.

- The Output Voltage and Current Waveform of Single phase full-bridge inverter with RLC Load are as follows:

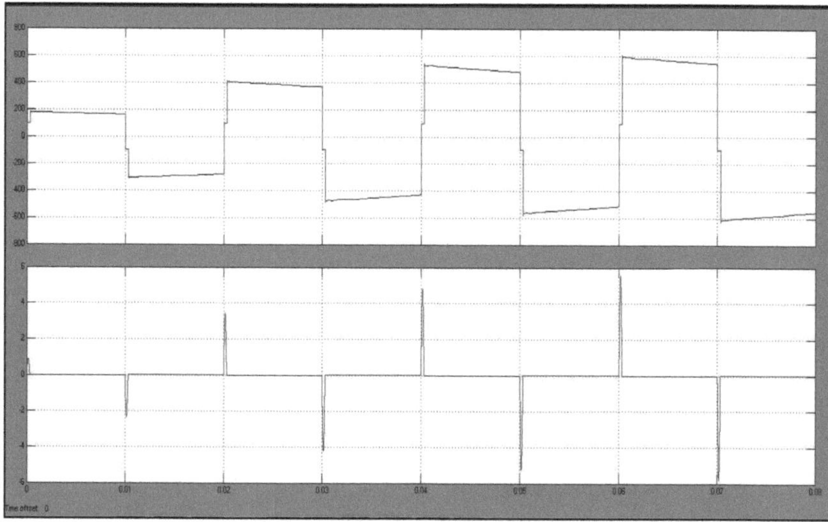

Figure 2.5

- For Figure 2.8,

X-axis defines Time Period and Y-axis defines Voltage Value and Current Value respectively

Chapter 3

Result

> FFT Analysis of Single Phase Inverters

i. Click on powergui then click on FFT Analysis

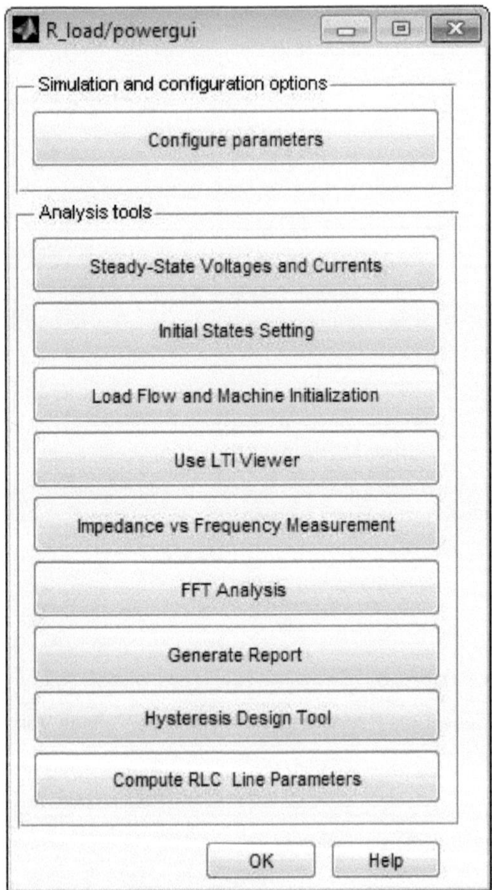

<u>Figure 3.1</u>

1. FFT Analysis of Single phase half-bridge inverter

> Single phase half-bridge inverter with R load.

The upper section of the figure shows the selected scope signal and the lower section of the figure shows its FFT analysis.

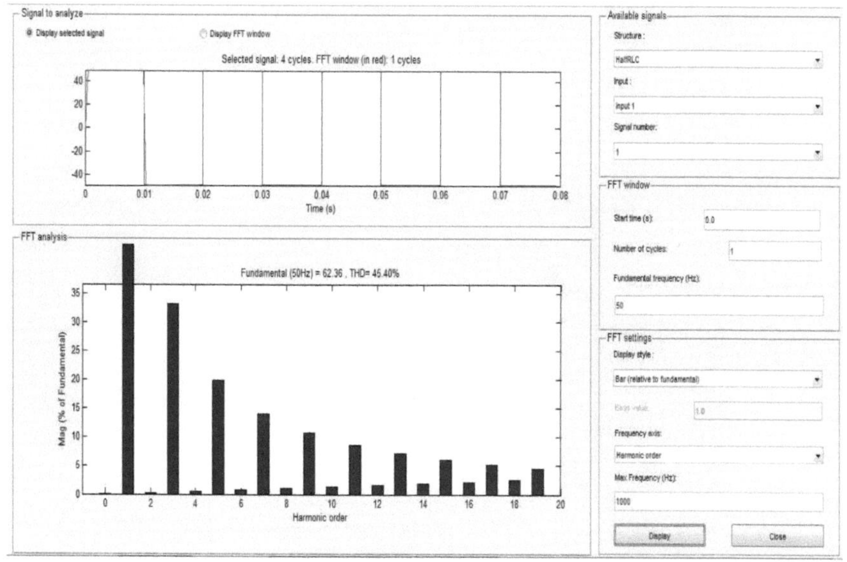

Figure 3.2

➢ Single phase half-bridge inverter with RL load

The upper section of the figure shows the selected scope signal and the lower section of the figure shows its FFT analysis.

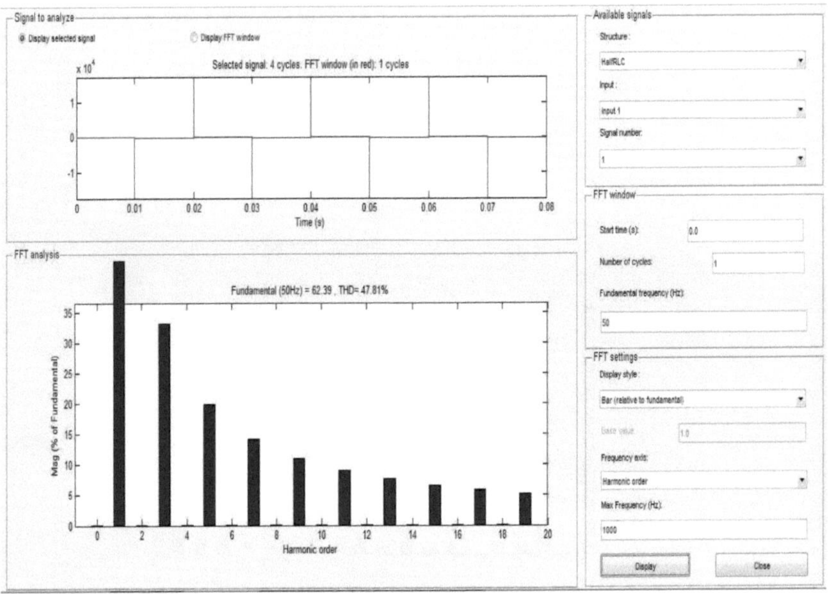

Figure 3.3

> Single phase half-bridge inverter with RLC load

The upper section of the figure shows the selected scope signal and the lower section of the figure shows its FFT analysis.

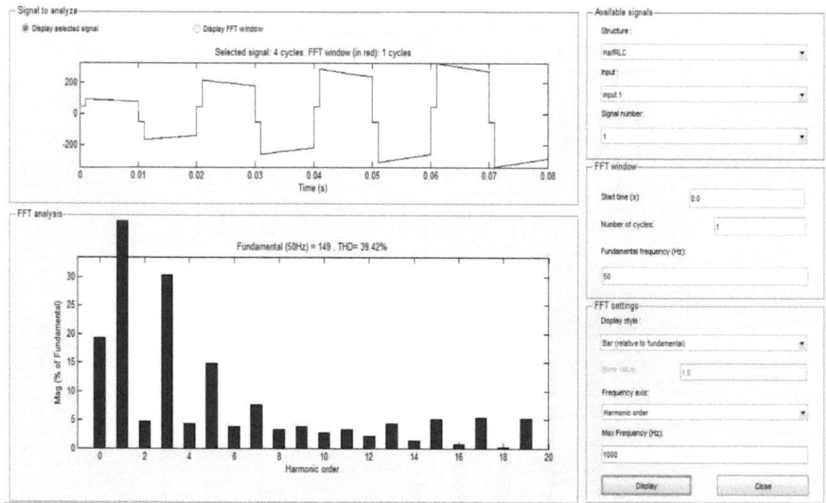

Figure 3.4

2. FFT Analysis of Single Phase full-bridge inverter

> Analysis of Single Phase full-bridge inverter with R load

The upper section of the figure shows the selected scope signal and the lower section of the figure shows its FFT analysis.

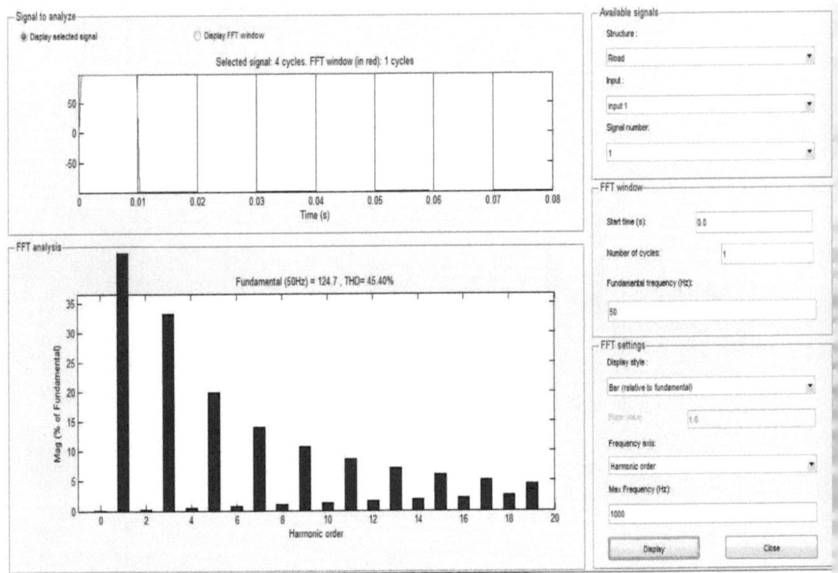

Figure 3.5

➤ Analysis of Single Phase full-bridge inverter with RL load

The upper section of the figure shows the selected scope signal and the lower section of the figure shows its FFT analysis.

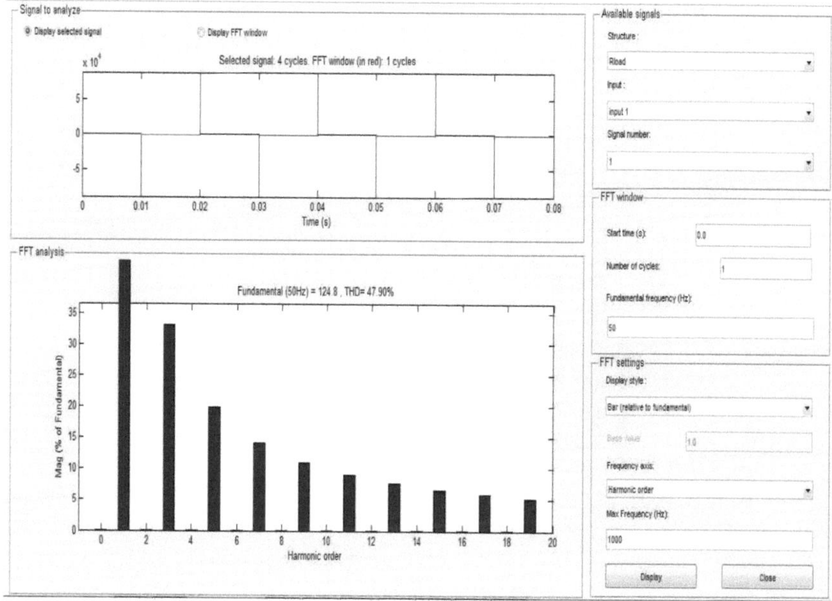

Figure 3.6

➤ Analysis of Single Phase full-bridge inverter with RLC load

The upper section of the figure shows the selected scope signal and the lower section of the figure shows its FFT analysis.

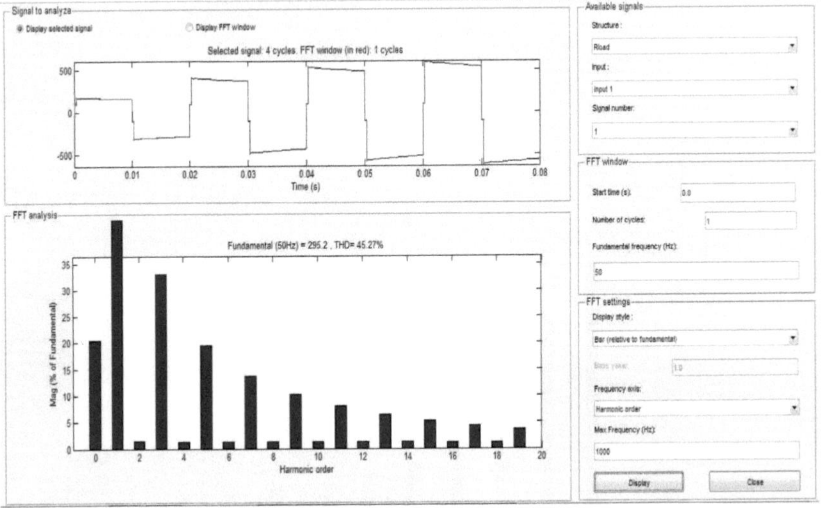

Figure 3.7

Chapter 4

Conclusion

We have studied about single phase inverters (IGBT) with different loads. The study is based on simulation results of these circuit using MATLAB.

Results of circuit is shown in Chapter 3 RESULT.

We have already studied about harmonics generated due to switching action of semiconductor devices i.e IGBT.As we have seen from the results that lower order harmonics are very prominent/dominant as compared to the higher order harmonics.

Further Scope

We have studied the Harmonic generated in single phase Inverter. These harmonics are injected in the supply system & deteriorates the power quality. Nowadays equipment's are very sophisticated in reference to power quality issues so this is to avoid malfunctioning or damage of expensive equipment's. Power quality should be good & for that harmonics in the system must be as prescribed by the standards.

We suggest different switching scheme in inverter for reducing the injection of harmonics.

References

> P. S. Bimbhra Publisher, Power Electronics ,Khanna Publishers (2012) , ISBN-13 9788174092793, ISBN-10 817409279X English.

> Muhammad H. Rashid , Power Electronics: Circuits, Devices, and Applications ,Edition reprint Publisher Pearson Education India, 2004 ISBN 8131702464, 9788131702468, 880 pages.

> MIT open-courseware, Power Electronics, Spring 2007.